ASSUMPTION AND MYTH IN PHYSICAL THEORY

BY

H. BONDI, F.R.S.

Professor of Applied Mathematics at King's College in the University of London

THE TARNER LECTURES
DELIVERED AT CAMBRIDGE
IN NOVEMBER 1965

CAMBRIDGE
AT THE UNIVERSITY PRESS
1967

CAMBRIDGE UNIVERSITY PRESS
Cambridge, New York, Melbourne, Madrid, Cape Town, Singapore,
São Paulo, Delhi

Cambridge University Press
The Edinburgh Building, Cambridge CB2 8RU, UK

Published in the United States of America by
Cambridge University Press, New York

www.cambridge.org
Information on this title: www.cambridge.org/9780521118064

First published 1967
This digitally printed version 2009

A catalogue record for this publication is available from the British Library

Library of Congress Catalogue Card Number: 67–21954

ISBN 978-0-521-04282-6 hardback
ISBN 978-0-521-11806-4 paperback

PREFACE

When I received the invitation to give the Tarner lectures I thought that they would be a good opportunity for expressing some of the accumulated prejudices and notions that have formed in my mind during quite a few years. A scientist really has a very limited number of occasions of speaking his mind. He can publish a paper, or talk about a paper at a scientific meeting. But it is one of the misfortunes of science that this is an exceedingly stylized type of presentation. A few years ago Sir Peter Medawar gave the first of a series of scientific talks on the BBC under the title, 'Is the scientific paper a fraud?' He finished with a strong affirmative, and I have always felt like applauding this conclusion, because the form and style one has to adopt in writing a scientific paper do a great deal of violence to any true history of one's thinking. The reader of the paper learns as little as may be of what the author's thoughts and intentions were. For what the author is supposed to do—and I think I am right in saying that the pure mathematicians are perhaps the guiltiest of the lot, but scientists in physics or chemistry are not much better—is to present his result in a manner as though it had been a sudden flash of inspiration. He is to give no hint of how he ever came to think about the problem. He then, having stated what his conclusions are, proceeds to prove them by theorems sufficiently rigorous to browbeat the reader into acquiescence, if not actual assent. The aim is to leave as disembodied, as impersonalized a piece of writing as anybody might be willing to read, knowing that others have to read it if they wish to know what has been achieved. The paper is very likely to tell the reader almost nothing about how the result was found. Often the derivation presented in the paper is about the fifth the author came across (I have not carried out a statistical survey but I think this is of about the right order of magnitude) and, as to why the author ever bothered to think about that particular problem at all, only a very daring author would breathe a hint of that.

The scientific paper is thus a very special form of presentation. This has certain advantages, particularly the advantages of brevity. It has the enormous disadvantage of making science—which in my view is the most human of all activities—totally impersonal, and of course makes it so necessary for scientists to go and meet each other and to talk about what they are not allowed to say in their papers. This is always very hard on the young people who have not got quite the chance to travel and see people at the other end of the Earth, as travel funds, however freely available to senior people, are always scarce for the younger scientist. I hope I have indicated enough to say what a strait-jacket we often feel the form of the paper is.

A second method of giving vent to one's ideas is in

the form of a book. A book is quite a nice object, but there is generally an awful lot of it, which is tough on the reader and even tougher on the writer. Moreover, a book usually, though not always, has to have some sort of subject or theme, and so a book is not really the best medium for an author who wishes to give free range to his rambling thoughts on one subject or the other as they come into his mind. The kind of lecture which I have been so kindly invited to give, and which now appears in book form, gives one a rare opportunity to allow the bees in one's bonnet to buzz even more noisily than usual.

The reader should therefore not expect a strongly coherent whole nor a set of exact solutions to precisely formulated questions, but rather an amalgam of unresolved questions, of personal opinions on teaching method, and of even more personal views on what I regard as some of the myths of science.

King's College London H. BONDI
November 1966

CONTENTS

I

THE LIMITS OF THEORY MAKING

The existence of scientific theories is one of the most significant, probably *the* most significant, aspect of the kind of thought that we call scientific, which has been demarcated for us so well by Popper. If I say a good deal here that reads just like a variety of foot-notes to Popper, then that is exactly what I intend it to be. I want to make some remarks about the by-ways of theory-making that are not necessarily com-prised under Popper's hat. As you will recollect—to give a very brief presentation for those who don't know about it—in Popper's view the act of making a theory is essentially an act of the imagination in which the scientist is naturally guided by the empirical (experimental and observational) knowledge of his day. The purpose of such a theory is in the first instance to comprehend, to encompass, what is al-ready known. But over and above this it is an essential requirement for a theory to stick out its neck, to make statements that can be tested by experiment and ob-servation. Now the very important point that Popper makes is that if experiment and observation go *against* the theory, then the theory has been disproved. But in no circumstances can we say that if the experi-ments or observations go *for* the theory then the theory has been proved. A theory is tested by experiment and observation and, if it has passed one test, then it is the task of the theory to make further forecasts—to go on, as it were, living dangerously—and to stick out its neck so that it can be tested again and again. Proof is completely absent from this picture, which I regard as a very accurate description of work in science; disproof plays a very large and vital part in it. Indeed, we can say that an essential piece of progress in science is disproof. Now this is the basic Popper doctrine, with which I wholly agree, given in a pocket version, but still, I think, containing all its essentials.

On the other hand, I think one must make a number of footnotes to this, some of which are common knowledge amongst all philosophers of science, and others are just my own particular by-ways of thinking. I may say that if in what comes I talk a lot of bad philosophy of science, then I will ask you to remember that I am not a philosopher, but a physicist, and if my physics is a bit wonky, please remember that I am the holder of a Chair in mathematics.

Now there are two of three points that emerge from this. One's whole basic instincts of justice and equity are likely to rebel against the case of the theory that has been tested time and time again, and then one day fails and is disproved. What is the position of such a theory? Does it just go on the scrap-heap and is forgotten? I think that here one must be very cautious and careful. If a theory has passed a considerable number of tests, then we know that there is a region of knowledge—empirical knowledge—adequately described by the theory. It is true that by refinement of measurement or by going to a different region we found the theory to be false. But that does not mean that the theory is invalid and useless within the context in which it was established. When an architect tries to build a house, then, as in the days of the ancient Egyptians, he makes plans on the assumption that the Earth is flat. This assumption has been disproved quite cogently, but nevertheless the method of working in some restricted area as though the Earth was flat continues to be carried out. The difference between the period before the theory of the flatness of the Earth was disproved and the present is simply that in the old days one would say, not merely: this is a good method of doing what I am about to do, but one would have said: this is the truth. There are, of course, debates, which I have never quite understood or followed, about whether truth is a word that has any place in science at all. My own inclination is that science has nothing to do with truth, but I am not sufficiently well versed in this to argue about it at any great length. Anyway, what you will have destroyed by the disproof of the flatness of the Earth is the attitude to it, that this is what really is the case. But that does not diminish the utility of continuing to employ, in a restricted way, a method that has been well tested.

In a slightly less ridiculous setting—a much less ridiculous setting—we can take the example of Newton's theory of gravitation. I think I am right in saying that no physical theory had ever been tested and checked and cross-checked as much and with as high accuracy as Newton's theory of gravitation in the first 220 or so years of its application. But all this did not save the theory from being disproved by certain discrepancies, which only turned out to be there when exceptionally high accuracy of measurement was used well over 200 years after the establishment of the theory. What the disproof of Newton's theory means is that we cannot claim it to be a perfect description of the way gravitation actually behaves. What we certainly know as a result of this enormous range of tests that have been carried out is that if we want to know the result of gravitational theory within certain well-known, well-established limits of accuracy we would be foolish to use any more complicated theory than Newton's. It is only when we want to go to these last refinements, or when we want to go into deeper explanations and investigations, that we abandon Newton's theory, and then we abandon it with some degree of confidence because we know that it does not quite fit the observational results. We certainly abandon it with much more confidence than we embrace anything else, but that is just by the way.

The next point I want to make is a purely political point, and it is connected with the progress of science.

I have spoken earlier of disproof as the essential agent of progress, but why can we disprove today what we couldn't disprove yesterday? The answer is that to-day we can carry out more accurate experiments relating to matters that were inaccessible yesterday. We can do so because of the progress of technology. And so a progressing technology is an absolutely essential condition for a progressing science. It is a peculiar disease of this country, I think, to feel that science sort of marches in front and that poor, dirty technology follows a long way behind. But the relation of science and technology is the relation of the chicken and the egg; you cannot have the one without the other. It is true that modern technology derives from modern science, but we would not have had any of modern science without modern technology. The enormous stream of discoveries at the end of the nineteenth century that gave us such insight as the discovery of electrons, discovery of X-rays, working with radio-activity, and all that, is entirely due to the fact that the technologists developed decent vacuum pumps; until you have decent vacuum pumps you cannot do any experiments of the kind required. Of course, X-rays are a splendid scientific discovery, but when this is properly used by technologists to make reliable, safe, accurate X-ray equipment, you can employ it to make progress in molecular biology. This is, as I was saying, a political point because it is a local prejudice that technology is in some sense second rate, but technology is just as much a precondition for science as science is a pre-condition for technology, a point also stressed by Popper.

I would like now to come to a point that I referred to earlier when I said that we would not use Newton's theory of gravitation, not only when we wanted to investigate matters with very high accuracy, but also when we wanted to think in greater depth. I feel that this question of depth is a terribly important one, and one on which I am myself unclear. Of course notions of depth differ in differing subjects; the pure mathematicians have very strong feelings on it, and I just cannot follow them there—I am ignorant of that subject. But in physics I think depth has a reasonably well-defined range of meanings. First of all, we regard something as deep if, and only if, it has a certain degree of universality. Universality is a very peculiar thing in science. It is not a scientific result to say that I enjoyed the steak I was served for lunch yesterday. It may be that the cook regards this as important, but I don't think the physicists would. And the reason for this is that anything we want to do must be repeatable, must be checkable, by others elsewhere. How far can we go with this? Checking, testing, all the like, is of course inherent in the definition of science, as I said earlier. But how far can we carry this assumption that what can be treated scientifically must be repeatable? I think we are here in a very interesting domain, a domain that (to be precise) becomes interesting in one connection only—but then I have my professional prejudices too—and that is in the direction of cosmology. If you have anything that depends upon an absolute world time—if you say that the universe has a definite age—then of course you must be aware of the possibility that this fundamental universal time enters your equations. And since the test, the checking of an experiment will always be later than the experiment itself, it follows that this variable is not repeatable; this variable cannot be given the same value as it had before. We are very far from being in a position even to specify an experiment where such an effect could show itself. It is true, though, that some very distinguished theories of cosmology suggest that as basic a quantity as the constant of gravitation depends on the absolute universal time monotonically. On this basis there is thus no possibility of strict repeatability. At what stage and to what extent this upsets the ability of science to deal with a problem of this kind is a very debatable point on which not only have different people expressed different opinions, but some of us (I must guiltily admit) have expressed different opinions on different occasions. This, then, is just a point, that the universality that we desire for any form of fundamental science is perhaps not as easy to attain as one sometimes thinks. Nevertheless, I tend to think that we would regard a physical theory as deep only if it has a good degree of universality. Wearing my hat as an astronomer, I can therefore say that any theory which solely deals with certain circumstances relating, shall we say, to peculiarities of our atmosphere, cannot be deep because we would not necessarily expect an atmosphere anywhere else to be just the same in distribution, in geographical obstacles it has to meet in its motion, in its source of energy, and so on. On the other hand, atomic physics automatically has an aura of depth, and so has anything relating to fundamental particles, because we have every reason to believe that atoms elsewhere and at other times are exactly just like atoms here and now; so that what we establish here and now has a great deal of universal validity, and hence of depth. Similarly, going back to Newton, the law of gravitation, which so often is called the law of universal gravitation (in itself indicating that one is dealing with something that one expects to be valid everywhere and always under any circumstances) is accordingly deep. This has proved on occasion a rather dangerous line of thought, or so at least it would seem to me. We have had, mainly in the last 50 years, but probably earlier too, various attempts to come to ultimate equations, to come to ultimate final complete statements, to theories unifying all that we know; and it does seem to me that this is a very dangerous tendency in the search for depth. And I say it is dangerous, not only because it has proved fruitless, but also because I find it personally repugnant. I have had these objections now for a good long time, even in the days when every respectable relativist was looking for an equation unifying all the forces that anybody had ever come across. It seems to me that there are two points about science that one must keep in mind. One, that I have already talked about, is its progressive character, and it is certainly a matter of experience that every time our experimental technique has taken a leap forward, we have found things totally unexpected and wholly unimagined before. I see no reason whatever to expect that future improvements in experimental technique will not have the same effect. If, then, we have a theory that is in some sense an open theory, then we can accommodate at least some new discoveries. The theory will retain its utility even after the discovery of new things—at least, many of our theories will. For an example of what I mean by an open theory, take Newton's second law of dynamics—that the rate of change of momentum of a body equals the force applied. It is a perfectly precise statement, but it leaves it entirely open to you to put in under the heading of 'force' any force so far discovered. And if you find some new kind of force there is no reason why you should not put that in. The theory has a ready-made place in it for putting in something new and unexpected. That does not mean that the theory cannot be disproved; indeed, we know that Newtonian dynamics has been superseded by relativistic dynamics. It does not mean that the theory does not say anything; everything that you can disprove says something. But nevertheless it is not a closed theory. More than that, I regard it as an essential of any scientific theory to have room for putting in what one does not know yet. We all come across people who are not very good at making up their minds and whose constant answer is: sorry, I can't make a decision on this; I am not yet in possession of all the facts. The scientist, of course, is somebody who is never in possession of all the facts because something new is likely to turn up at any moment. You must always be prepared to deal with the unknown. It is an essential part of science that you should be able to describe matters in a way where you can say something without knowing everything. I have heard it said—I don't know whether there is any evidence for it or whether it is just invented—that when Newton first discussed the attraction of the Earth on the moon, his opponents said it was ridiculous to attempt to make a theory of the attraction of the Earth on the moon when one knew next to nothing about the interior of the Earth. It is an eminently reasonable point of view. But the test of science is not whether you are reasonable—there would not be much of physics if that was the case—the test is whether it works. And the great point about Newton's theory of gravitation was that it worked, that you could actually say something about the motion of the moon without knowing very much about the constitution of the Earth. You must remember, of course, that the opponents were not stupid and that there is something in what they were saying. For example, you cannot discuss the secular acceleration of the moon, which is due to the friction of the tides on the Earth—which has turned out to be mainly due to the tides of the solid Earth, I understand—without knowing how imperfect the elasticity of the material of the Earth is. So it *is* true that you can't say absolutely everything about the motion of the moon until you understand the constitution of the Earth completely. But it is an essential property of science that you should be able to make statements that say a great deal about the motion of the moon without knowing a great deal about what the Earth consists of. All science is full of statements where you put the best face on your ignorance, where you say: true enough, we know awfully little about this, but more or less irrespective of the stuff that we don't know about, we can make certain useful deductions.

Now, my view is that any theory which pretends to comprehend everything breaks down on this point. It will be a uselessly rigid theory because it won't have room to put new discoveries, it won't have a place into which to put new things. Of course new discoveries may always upset some theory and wreck it completely, but at least we ought so to shape our theories that new discoveries won't upset *every* theory we have and for that purpose we must have plenty of *open* theories.

The kind of heresy I am storming against was much more popular 30 years ago, particularly when both Eddington and Milne put forward theories which they thought were all-inclusive. But even nowadays a physicist as distinguished as Heisenberg tries to write down the 'world equation' which, he hopes, says everything. Of course there are so many other things you can say against this attitude—that an equation that says everything says nothing, because if the enormous variety of things that we see in this remarkably variegated world all spring from one equation, then the way from the equation to the things that we see must be awfully long and very difficult to deal with. In a sense, however, this is a criticism of all fundamental work. I hope that you do not think that I wish to condemn root and branch all search for depth; very far from it. It is a very good thing to do, within reason, but one should in particular try not to eliminate entirely the openness of the theory so that it can in some respects be accommodating. In other respects it must be rigid, because a theory that is not rigid enough to be disproved is just a flabby bit of talk. A theory is scientific only if it can be disproved. But the moment you try to cover absolutely everything the chances are that you cover nothing.

This is putting rather an unkind interpretation on work like that of Eddington, of Milne, and of Heisenberg. In particular Eddington used to argue that if we just followed the dictates of our thought, the structure of our minds, we would necessarily discover the whole of physics. In fact, his idea was very much that epistemology is *the* tool for finding new laws of physics. Fortunately, I think, there are very few adherents of this view now. Unfortunately, a very useful aspect of it has been lost sight of. I am certainly not a psychologist, but my impression is that looking for the inherent structure of our minds (i.e. the necessary ways of thought) is a useless thing, because our intelligence consists essentially of the flexibility and adaptability of our minds. It is just the fact that they can be adapted to everything that makes our minds useful. On the other hand, we must remember when and how our minds are formed. They are formed very much in the first few years of life by contact with all the things we play with in those days. We learn an enormous amount of physics in the first two or three years of our lives. In fact, we learn this bit of physics much more thoroughly and unforgettably than anything we learn later. (Probably also painlessly, though this may only be an effect of infantile amnesia.) It has seemed to me that these extremely important aspects of physics which we discover in the first few years of our lives may in fact contain rather more information than we commonly ascribe to them. We very often find new results, new insights, by doing very complicated experiments. And it is only much later that it is realized that we could have got the same insight through rather simpler experiments. Eventually the crucial experiments can be performed at school. But I don't myself think that quite enough effort goes into exploring this. It may be that an awful lot of high-grade physics is already implicit in the simple physics we learn in the first three years of our lives. It would seem to me to be entirely tenable if it were claimed that you could deduce the atomic and quantal structure of matter from the fact that there are solid bodies. I have not seen this done in any rigour, but I think it would be a very worthwhile enterprise if one could show that a good deal of the high-grade physics, derived normally from rather refined observations, is already inherent in what is common knowledge. Perhaps we should try to look at Eddington's attempt to see far in physics epistemologically not in his sense, but in the sense that much deep physics may be hidden, not in the basic structure of our minds, but in what entered and formed our minds at an early age in a rather simple way. Then it may be that we could get a great deal further. But not much has been done in this direction.

I have been very much interested in ways of making special relativity simpler to understand, and I have always said that my ultimate aim is to get special relativity into the primary school syllabus. What one wants to do in order to achieve this has also seemed to me quite simple. If somebody could devise a cheap and safe toy which is an accelerator fast enough to show relativistic effects so that children of 5, 6 or 7 could play with it, then special relativity would be regarded as obvious and a suitable subject for the primary school curriculum. I have heard it said that the best way for the developed nations to help the industrially underdeveloped ones is by flooding them with cheap mechanical toys so that the children there grow up with mechanical devices and get a liking and an interest for them. If we could do the same in physics I think we would see a great deal further. To me it is a very great physicist who can show by using a torch battery what has been proved previously by using an enormous accelerator. This kind of physics, I think, needs more support, needs more pushing.

I want to conclude this chapter with a few remarks about the way our theories have been limited, not only by the requirement to say something without saying everything, but also historically. I think that all the great achievements in the development of physics have carried with them prejudices and myths, myths that have sometimes been helpful and sometimes been extremely harmful to the further development. My pet object in this direction is Maxwell's electromagnetic theory, a fantastically beautiful, simple, comprehensive theory which marked one of the really great advances in physics just over 100 years ago. But it carried with it a great deal of baggage, and some of this baggage has made, and I think still is making, work in other directions of physics somewhat more difficult than it need otherwise have been. Because Maxwell, following Faraday, chose to make his theory a field theory, the myth entered physics that *all* good theories were field theories. Now, I do not wish to say that field theories are wicked or evil or anything like that, but that minds have been closed for perhaps rather longer than was necessarily desirable to the possibility of considering other kinds of theories. I regarded it as a very great step forward when Wheeler and Feynman some 20 years ago managed to formulate Maxwell's theory in a particle action form. This helped to free science from a myth.

Some myths arise because a great scientist, looking at things in a certain way, advanced considerably. Others are then liable to say: looking at things in his way is the prescription for getting further everywhere and at all times. This attitude can be very strong. I want to remind you just how kind Nature has been to us in relation to electromagnetism. In particular I want to remind you of the absurd fluke that has made electromagnetic theory so much more intelligible than it might otherwise be, namely Ohm's law. Here the fact is that many ordinary materials satisfy a ridiculously simple rule that does not follow in any very direct way from any obvious assumptions. What Ohm's law really means is that in many materials the *velocity* of the charge carriers is proportional to force, whereas we have known since Newton that it is *acceleration* that goes with force. Ohm's law gives us an expression for the dissipation of energy of a simplicity that is not, to my knowledge, equalled anywhere else. If you think of the complication of hydrodynamics, or if you try to make a dissipatory gravitation theory, as I have tried to do on occasions, you will appreciate what an absurd fluke it is to have such a simple formula that always leads to the dissipation of energy, never to its generation, that is linear, and which applies to a wide range of materials. Again, what helped so much in the development of the electromagnetic theory has probably spoiled us; we expect the same simplicity elsewhere, and we are very disappointed when we don't find it. So here again we have a limit on theory-making; the new theories may not always operate in quite as simple a way as the past ones.

2

RELATIVITY: ITS MYTHS AND PRE-SUPPOSITIONS

Relativity is a subject on which one can spread oneself very much, because it contains a good deal of ordinary common-sense, and around it there has grown an extraordinary accretion of science fiction and nonsense and myth, and above all of mystification. I am very much under the impression that popular writers, for quite a few years after the origin of relativity, thought that their books would sell better the less intelligible they were. And this led to a unique succession of efforts, each better than its predecessor. I became interested in this subject some years ago—meaning by this subject not the obvious special relativity which every physicist knows, but the subject of trying to strip the overlying nonsense and mystery off and trying to see what simple, straightforward notions remain.

Let me start by speaking of the very idea of relativity. I would be stupid if I were to say that science must follow certain rules and regulations for all time. On the other hand, I have a strong inclination to believe that some kind of repeatability of experiments is rather necessary to any growth of science at any time. I am not even attempting to specify in any detail what this must or should be. But any assumption of any form of repeatability implies some independence of at least some branches of physics from some circumstances, such as location in time, location in space or state of motion. Obviously the greater this range of independence from all these various things, the easier it is to do physics. And at the present stage (by which I roughly mean from the days of Galileo to the present day, both included) we have a pretty wide range. In particular let me suggest that, for what we might call laboratory physics, at the least, there is independence of location in space, independence of location in time and also a limited amount of independence from the state of motion. I think it is often worthwhile to remember just how extremely different this post-Copernican view is from the pre-Copernican. As long as it was thought that our Earth was the centre of the universe, it followed that everything that took place here was completely special and thus entirely atypical. Accordingly, what was found to be the case here could not be assumed to hold anywhere else, because of the very special property of 'here' being in the centre of things. It is very good for one's ego to think that one is in a completely special and highly selected position but it is an awful nuisance when you try to do science. As you know, the work of Copernicus and Galileo and Newton saved us from this presumption of being all that special, and since then we have always had the idea that, by and large, things elsewhere aren't all that different from what they are like here. I think this independence of location in space is an idea that, even in those days, was quite easy to grasp. What was difficult to understand then about Galileo's and Newton's work, and what is difficult to understand now, is the *limited* independence from the state of motion. If one were a philosopher completely ignorant of physics one might say that one of two alternatives represented a reasonable description of the dependence of physics on motion. You could either say that there is one specially chosen specially selected state of motion which we may call rest, and that all the others are different. That would be a perfectly sensible, clear-cut statement. Or you could take the view that the state of motion doesn't matter for physics. As you know, we don't have either of these situations. We have the very peculiar situation that there is a *set* of states of motion, namely inertial motions, such that there is equivalence between different people moving in different ways inertially, and inequivalence between any of these and the rest. Since the days of Galileo and Newton we have regarded velocity as relative but acceleration as absolute. This is difficult to understand, but there doesn't seem to be any simple way out.

I might still remind you how closely interwoven this independence from velocity is with the whole Copernican picture. The first reaction of the man in the street 300 or 350 years ago to the Copernican theory would have been 'What rubbish to suggest that the Earth is moving—I can't feel it to be moving.' This is, as a first reaction, very sensible indeed. It is only because of this independence of physics from velocity—together with some much more high-brow stuff referring to gravitation (but that only in a secondary way)—that makes the whole Copernican picture tolerable at all. Thus we have arrived at this outlook of Galileo and Newton, that there is a selected set of states of motion equivalent to each other but inequivalent to other states of motion. There is nothing more difficult to grasp about relativity than Newtonian relativity—that there are inertial observers, that one of them is as good as any of the others, but that acceleration is something quite different.

I have been at pains to stress this point because I don't think it is pointed out in all presentations that the really difficult step is the step taken by Newton and Galileo, and not the step taken by Einstein. Because we learn Newtonian physics at a stage of life when our receptive power is very high and our critical power not over-great, we tend to swallow these very difficult ideas without much questioning, and then think that the difficulty lies in what we happen to learn later when our receptive powers are poorer and our critical powers a little greater. But we find this difficulty in the Einsteinian picture only because we have not digested it sufficiently in the Newtonian picture. Indeed, I would tend to say that Einstein was a return to Newton. For after all in Newtonian dynamics, and in particular in the first law of dynamics, you have this clear selection of inertial states of movement from all other ways of moving. It is an immediate logical deduction from the first law that all inertial observers are equivalent as far as Newton's first law goes. By plausible inference this equivalence extends to the other laws of motion, and therefore to the whole of dynamics. One does not at this stage assert that the equivalence of inertial observers is necessarily confined to dynamics. This restriction was a later nineteenth-century addition to the picture coming in through the then prevailing views on the propagation of light. I think we still have a tendency to teach people relativity by very carefully and deliberately first teaching them the nineteenth-century point of view, in which there was no equivalence between inertial observers as far as light was concerned because in that view velocity relative to the ether mattered; and afterwards, when we have succeeded in teaching them the nineteenth-century point of view, we then say: ah, yes, but that's all wrong. I see no sound reason why one should ever go through these convolutions. If you start with Newtonian dynamics, the whole notion of equivalent inertial observers follows immediately, and also with it the fact that the relative velocity of any two inertial observers is constant in time. This is Newton's principle of relativity.

Where the crucial step lies at that stage is in the uncritical assumption of an absolute time. This is not explicitly part of the contents of Newton's laws of motion, though of course it is explicitly part of

Newton's picture, and if at that stage one could say, 'For the moment we can assume a universal time, but remember that we add this to Newton's first law', then I think there should be no great difficulty in making the transition to special relativity later.

All Einstein did, then, from this point of view, is to say: it follows strictly logically from Newton's first law that there exists a set of observers to be called inertial observers such that any one of these finds Newton's first law to be correct and nobody else finds Newton's first law to be correct, and that the relative velocity of any two of these inertial observers is constant in time. And the obvious inverse of this is that all these inertial observers are equivalent as far as Newton's first law goes, because that is how we have defined them. In Newtonian dynamics the step is taken of widening this equivalence to *all dynamical phenomena*. What Einstein did was to widen this equivalence to *all phenomena*. What in fact one says, then, is this: we have arrived at Newtonian relativity from Newton's laws of motion, and with a slight possibility of argument we can say that it applies to all purely dynamical phenomena. You can then apply a little bit of criticism to this and say that there is no such thing as a purely dynamical phenomenon, because in anything that you think about in physics, other branches enter too; even in the prototype of all dynamical phenomena, the collision between two billiard balls, you rely on the material properties of billiard balls, which as we know are due to complicated quantal electromagnetic forces. And indeed there cannot be a purely dynamical phenomenon. Therefore if Newtonian relativity does not extend to the rest of physics, then it is an empty statement. It applies to a class of phenomena, the purely dynamical ones, which is an empty class. If, on the other hand, we say that Newton's principle of relativity is something that corresponds to our experience in some measure, then, as a self-consistent hypothesis, let us assume that it applies to *the whole of physics*. This assumption is Einstein's principle of relativity. Clearly this is purely an hypothesis and must be tested by experiment.

It so happens that Einstein's principle of relativity on the equivalence of inertial observers is one of those few simple statements in physics from which it is readily possible to derive consequences that can be tested by experiment and observation. What has bedevilled this issue in text-books is the undue prominence given to the Michaelson–Morley experiment. There are curious historical points about this—I am not a historian by any means, but I have been told (and I believe this to be true) that Einstein said that at the time he wrote his basic paper on relativity (1905) he had never heard of the experiment. Later on when it was decided to reprint various essays on relativity it was decided by the publishers (with the advice of somebody) to start in the middle of one of Lorentz's essays. The first part that was included happened to be the Michaelson–Morley experiment.

For this reason since then everybody, or nearly everybody, has felt obliged to start in the same way. And what a complicated start it is! First you have got to explain the picture of ideas of the nineteenth century, which we now know to be inapplicable, out of which state of mind grew the desire to make this experiment. Then you have to say that it didn't have the result that was expected; therefore something must be wrong in what we have just taught you. And into all this goes a good deal of description of an awkward, difficult, experimental technique, and a little reference, perhaps, to the doubts that have arisen from time to time about this experiment. All this hasn't helped the understanding of relativity very much. What one surely wants to do is to take Einstein's *principle* of relativity and try and deduce from this principle easily observable results—which is not difficult. And then you can go forward.

The next thing that has bedevilled special relativity is a somewhat awkward attitude towards the measurement of distance. It has always bothered me a great deal—I don't think all my colleagues are quite so bothered—that it is frequently stated in the books that for relativity you require a rigid ruler. However, since it emerges from relativity that sound must not be faster than light, and since in a perfectly rigid ruler the velocity of sound is infinite, whereas the velocity of light is finite, you have an inconsistency which has always caused me no end of bother. I now think that we need not consider this particularly difficult instrument, the rigid ruler. What we do need to consider is the fact that Nature has very kindly given us a large set of identical clocks. We know that the frequencies of identical atoms—frequencies of particular lines of particular atoms—are well defined; we know that with certain precautions, like choosing the right isotope and so on, these atoms are identical in the sense of being indistinguishable—which is jolly useful—and that therefore we have simple time-keeping devices (simple conceptually, I mean) which do not seem to me to conflict in any way with the principle of relativity.

What do we do about distance? And here I feel that not nearly enough has been said about the deep debt of gratitude that we owe to Milne, who, 35 years ago in his work on cosmology, introduced the notion of the radar method of measuring distance. I have a saying that all good physics is potential engineering, and I think in few cases can this have shown itself as well as in Milne's attempt to use radar—that is to say, time measurement—for the measurement of distance. It is, of course, perfectly true that in his particular context there are certain difficulties about this. If we were to use radar on the cosmic scale we would have to be very patient indeed to wait a few hundred million years before we got signals back. But it is true that at present our best knowledge of the distance to the moon and our best knowledge of the size of the solar system is indeed derived by radar. So that this is a method of measuring distance which is not only conceptually simple, which not only means that a clock is quite sufficient—you don't need a rigid ruler—but also is one that is used in practice.

What this measure depends on very much are certain properties of light, and the basic property can be expressed, as I like to do it, in a completely non-metrical way: there is never any overtaking of light by light in empty space. What I mean by this, in a more low-brow way, is that in a vacuum all light has the same velocity; there is no dispersion by colour; there is no dispersion by intensity; there is no dispersion by the velocity of the source. Of course, this is also a hypothesis and no more—you can never make such sweeping statements and say they are straightforward deductions from experiment—but it is a hypothesis which has stood a large number of tests rather well. And one of these tests, as you probably know, concerns the light from spectroscopic double stars where the maximum red shift of the light of one source occurs just at the same time as the maximum blue shift of the light from the other, showing that, in spite of the different velocities of the sources, the light travels from the stars to us at the same speed. Of course always in physics we can explain away any experimental result in such ways as one likes by sufficiently complex hypotheses. But certainly the simplest way of interpreting this observation is to say that it is a test of the hypothesis that there is no overtaking of light by light in the vacuum. On this basis we can go forward. On this basis the distance measurement by radar is unique—it doesn't matter what kind of electromagnetic waves are used. If you have also (this is rather important) two bodies at the same time at the same place, but moving with very different velocities, and you illuminate them with your radar beam at the moment when they are together, then the scattered light from both of them will arrive at you at the same time, because, irrespective of their velocity, the light from both will travel together, there being no overtaking of light by light in empty space. Thus we will find the same distance for two bodies that were at the same point, in spite of the fact that they had a relative velocity.

A beautiful feature of using radar to determine distance is that all the talk about the velocity of light dissolves into nothing. You then measure distance by time, and your unit of distance is the light-year or the light-second or the light-millimicrosecond, whichever you like to choose. But there is no question of what the velocity of light is—it is *one*. Of course light takes a year to cover a light-year—if it took a different time we would not call the distance a light-year. Any attempt to measure the velocity of light is therefore not an attempt at measuring the velocity of light but an attempt at ascertaining the length of the standard metre in Paris in terms of light units. Any question about the alteration in the length of the standard metre is quite simply a question for solid-state physics—that is the branch of physics that tells you whether the length of something determined by inter-atomic forces changes in certain circumstances or doesn't change—this has nothing to do with us now; it is a much more complicated branch of physics than relativity.

I have discovered that what I call the *k*-calculus is still unknown in educated mathematical circles. In uneducated ones, of course, it is well known. And I propose, therefore, to talk a little bit about it here, partly because I enjoy it, partly because I hope to put those of you who don't know it into a state where you can also easily derive the consequences of relativity. You will then see where it leads you and see that Einstein's principle of relativity gives perfectly reasonable results capable of being tested by experiment and observation. The increase of mass with velocity, the equivalence of mass and energy, and the time dilatation have been checked with very great precision in countless experiments in accelerators (which indeed would not function if our formula for the increase of mass with velocity did not hold accurately) and innumerable observations on cosmic rays. These are the true empirical foundations of the theory, and not the Michaelson–Morley experiments and others of that period, which are even nowadays difficult experiments in which no great accuracy can be obtained. This is hence a way of destroying another old myth, namely that you must understand the Lorentz transformations in order to understand relativity. Now I would be the last person to say that

Lorentz transformations aren't useful—they are jolly useful—but it is not necessary to confine the class of those who can follow special relativity to the class of those sufficiently familiar with co-ordinate systems to understand the distinguishing features of the Lorentz transformations. It also seems to me nonsense that somebody should learn about co-ordinates for the sole purpose of being taught about Lorentz transformations. I don't think that is a very effective way of teaching, and it is not, I submit, the easiest way of teaching relativity. Of course it is perfectly equivalent to what I call the *k*-calculus, but you have to be reasonably high-brow to see the equivalence, whereas I think that the *k*-calculus approach does give you the essentials in a much simpler way.

The first great simplification arises—and this is in common with all old presentations of relativity right back to Einstein's trains—because one considers only one spatial dimension. A very large number of the features of special relativity show themselves with one spatial dimension, which is very good and very convenient. As you may know, I work in general relativity where one has to think about curved, four-dimensional spaces, and sometimes outsiders ask me how I can imagine such spaces. My answer is that a mathematician is distinguished from other people by the fact that he realizes that the human brain is not adequate to think in *three* dimensions. He has therefore evolved a mathematical apparatus which enables him to work in three dimensions without having to *imagine* three dimensions. This apparatus happens to be powerful enough to work in four dimensions as well but obviates equally the need to imagine four dimensions.

In continuing this approach to relativity, let me first make another attack on the customary use of the experiment of Michaelson and Morley. Suppose you were a modern physicist charged with the task of checking the Michaelson–Morley experiment. For those of you who have not been maltreated with it let me just mention that this is an experiment to compare the velocity of light in two different directions. A beam of light (Fig. 1) from one source is divided by a half silvered mirror into two beams at right angles to each other. The two beams are reflected back by two mirrors to the half-silvered mirror, so that an emergent beam contains light from both, and thus you get certain interference fringes. You now either turn the apparatus round or rely on the motion of the earth to turn it round, and you see whether your interference fringes move. This would be the case if the velocity of light depended on direction since this would affect the relative phase of light received from the beams.

Now suppose you were a modern physicist trying to repeat this experiment. You would be as well aware as your predecessors of the fact that it was essential to the experiment to keep the distance to the mirrors constant during the turning round of the apparatus. But of course you would use modern technology and not nineteenth-century technology to achieve this. You would not rely entirely on steel and concrete, but you might well prefer to have a built-in distance-measuring device and then use servo-motors

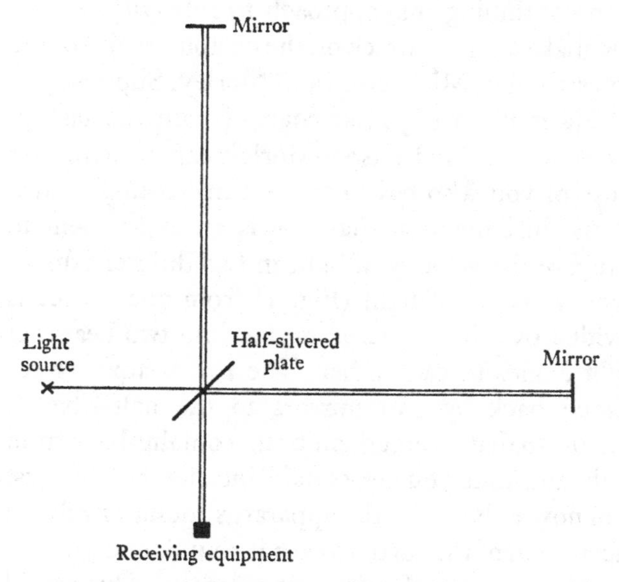

Fig. 1. The principle of the Michaelson–Morley experiment

to move the mirrors to keep them at the same distance. And surely the distance measuring device you would choose would use light-beams and interference measurements. And when you have decided to keep the mirrors at a distance so that the interference fringes don't move, then it seems scarcely worth performing the experiment to see whether they do move. There is a little more subtlety in it than appears in what I have been saying because there are different beams involved, but the result is essentially the same. So with our modern outlook and our modern technology the Michaelson–Morley experiment is a mere tautology. This is in no way a criticism of Michaelson and Morley, for their experiment was instrumental in establishing our modern outlook, but only of those who continue to base their explanation of relativity on a line of thought quite inappropriate to current thinking.

After this lengthy introduction, I am now ready to start on deriving the consequences of the principle of relativity by means of the *k*-calculus. The basic idea of the *k*-calculus is to consider a one-dimensional set of inertial observers, that is to say a set of observers all moving along one straight line. I can accordingly depict them in a space-time diagram using only two dimensions (Fig. 2), the vertical one representing some measure of time, and the horizontal one some measure of space. *Note, however, that these diagrams are in no way intended to be used for measuring, but only as a rough guide to memory to indicate who is where and when.* The observer A drawing the diagram is represented by the time axis, as he is always at the origin of the space co-ordinate. Any other inertial observer B moves relative to A with constant velocity and so we assume that we can depict him by another straight line. I wish to stress that a considerable amount of assumption goes into a simple diagram of this kind, though much of it relates only to the diagrammatic presentation and not to the theory behind it. I furthermore assume that both inertial observers are equipped with 'identical clocks'. You realize how heavily I lean here on quantum theory; but

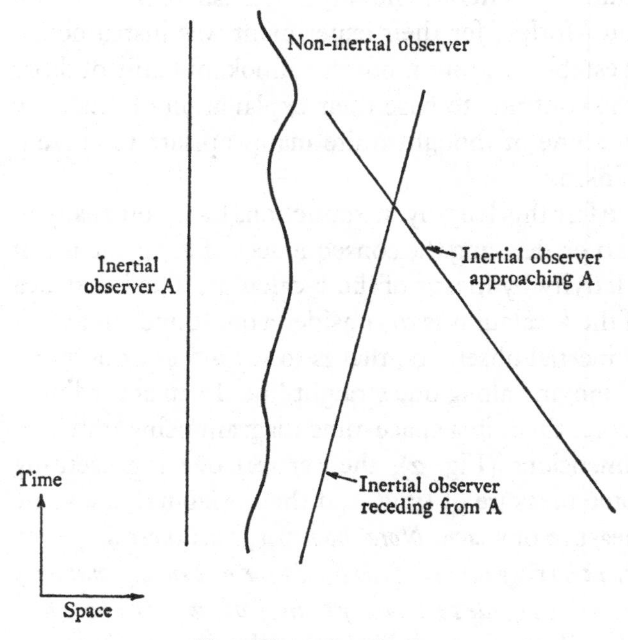

Fig. 2. Space-time diagram

for quantum theory we could not be certain that identical clocks exist. Thanks to quantum theory I can assume the existence of identical atoms controlling the clocks through their proper frequencies. I rely on quantum theory, I rely on the properties of light already mentioned, and I am also going to rely on the following rather recondite assumption: we suppose (Fig. 3) one of these observers

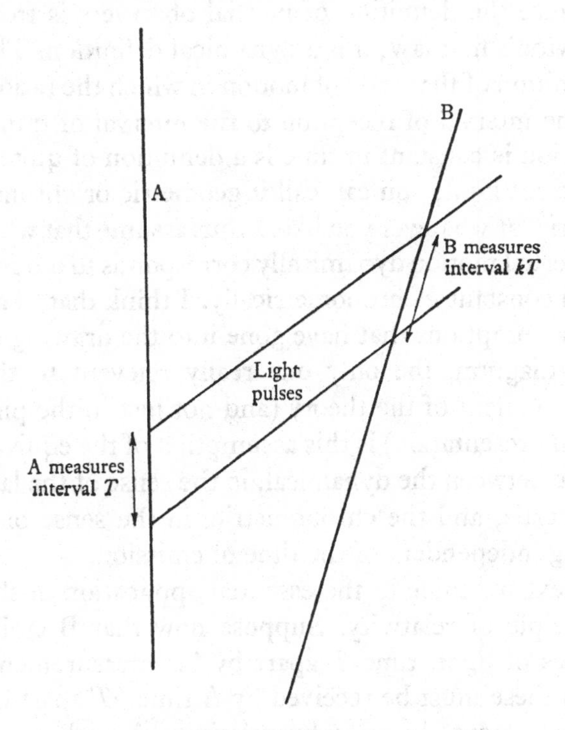

Fig. 3. The k-factor

(say A) to emit two light signals in succession; we suppose that by his clock there is an interval T between the emission of these two light signals. We now call kT the interval between the reception of these two by observer B, as measured by B's clock. We now assume that if A and B are inertial observers then k

is constant in time and independent of T. You realize that this is a rather far-reaching assumption, because the definition of inertial observers is from Newton's first law; it is a dynamical definition. The definition of that state of motion in which the ratio k of the interval of reception to the interval of transmission is constant in time is a definition of quite a different kind; you can call it geometric or chronometric, or whatever you like. I thus assume that what is inertial motion dynamically corresponds to motion with constant k chronometrically. I think that of all the assumptions that have gone into the drawing of this diagram, the only one really relevant to the development of the theory (and not just to the pictorial presentation) is this assumption of the equivalence between the dynamical, in the sense of the law of inertia, and the chronometric, in the sense of k being independent of the time of emission.

Next we come to the essential application of the principle of relativity. Suppose now that B emits pulses of light, time T apart by his measurement, then these must be received by A time kT apart by measurement, because by relativity A and B are equivalent. If the k factor were not the same one way round as the other way then there would be no equivalence between A and B; we could say one was more at rest 'really' and the other one less at rest 'really'. Let me remind you here that if you did this kind of work in the theory of the propagation of sound where there is a medium of sound propagation, namely the air, the k factor would strongly depend not only on the relative velocity of the transmitter and the receiver, but also on the velocity of each relative to the air. You get a different k factor with the transmitter of sound at rest and the receiver moving through the air, than with the receiver at rest and the transmitter of sound moving through the air. And by 'at rest' I mean at rest relative to the medium of sound propagation. You see all the nineteenth-century troubles hidden in this, because then it was thought that the propagation of light was much like the propagation of sound, only with a different medium. On the other hand, we now decry any such likeness since there is no medium of light propagation that is physically evident, whereas the air as the medium of sound propagation is physically observable in many other ways as well.

It is this point that the factor k is the same both ways round which is the crucial application of the principle of relativity. It is not obvious to me that there is any other place where it is used, but at this point it is vital.

The factor k is clearly characteristic of the relative velocity of A and B. This relative velocity can easily be worked out. We again use a diagram (Fig. 4) for pictorial representation. Suppose A and B exchange light signals when they pass each other. Since they are then zero distance apart, this exchange takes no time. A time T later by his own clock, A sends out another light signal. This will be received by B, according to our previous discussion, time kT by B's clock after his receipt of A's first signal. If B immediately responds by sending out a light signal, then this

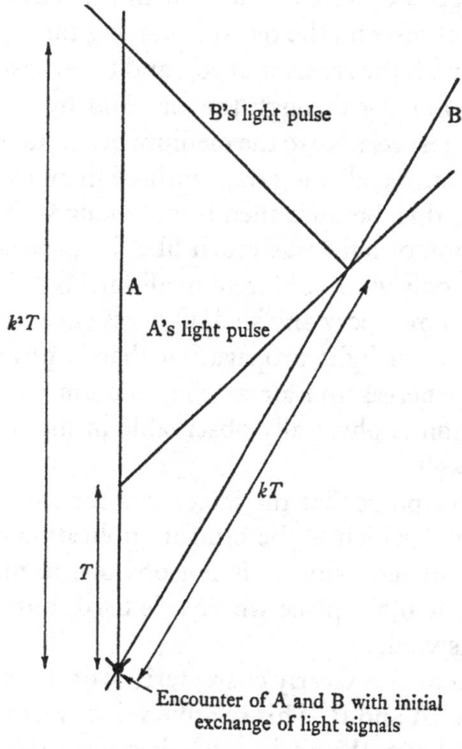

Fig. 4. Finding relative velocities from the k-factor

will thus be time kT after his first response, and will hence be received by A at time $k(kT) = k^2T$ (on A's clock) after the initial interchange. Thus in A's reckoning, the second light signal took time $(k^2 - 1)\,T$

for the round trip ABA. For the single journey the time taken is half of this, i.e. $\frac{1}{2}(k^2-1)T$, since the velocity of light is the same in one direction as in the other, being equal to one in both cases. Thus by the radar definition of distance $\frac{1}{2}(k^2-1)T$ is B's distance from A at the moment of reflection. What time does A *assign* to this moment of reflection? (And please realize that this is now something complicated.) For the first occasion in all these arguments one speaks of an *assignment* of time at a point other than where the observer and his clock are. We speak of *measurement* of time at one observer by that observer's clock —that is simple enough. But now we are assigning time at B by the clock of A. Again, since the velocity of light is 1 in each direction, the time that A assigns must be halfway between his emission and his reception of the light signal, which is therefore $\frac{1}{2}(k^2+1)T$ after the encounter of A and B. Hence A finds that B has changed his distance by $\frac{1}{2}(k^2-1)T$ in time $\frac{1}{2}(k^2+1)T$, and his velocity, v, is accordingly the ratio of the two expressions, namely

$$v = (k^2-1)/(k^2+1). \qquad (1)$$

This is a simple result connecting velocity and k factor.

Let me just do one or two other little things to show you just how easy the k-calculus is. Let us take 3 collinear observers, A, B and C (Fig. 5). There will clearly be a k factor k_{AB} between A and B, and similarly a k factor between B and C, k_{BC}. What is the k factor from the first to the third, i.e. k_{AC}? Again consider light signals, and remember that there is no overtaking of light by light. Observer A transmits two light signals time T apart by his clock. They are

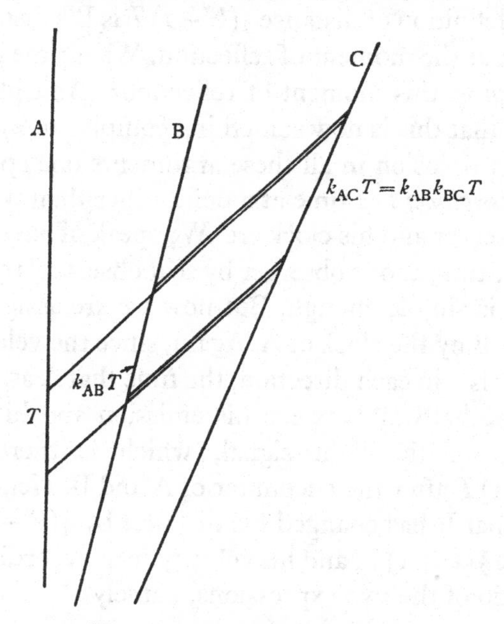

Fig. 5. The composition of velocities

seen by B separated by an interval (on B's own clock) of k_{AB} T, by the definition of the k factor. Similarly C will measure on his clock an interval k_{AC} T between the reception of A's two light signals. If now B transmits light signals towards C the moment he sees A's signals, then the light from B will travel with the light from A (no overtaking of light by light) and it will be received by C simultaneously with the light from A. Now the interval of transmission by B is $k_{AB}T$. Therefore the interval of reception by C is k_{BC} times as great and hence equals $k_{BC}k_{AB}T$. But at the same moment C will receive the light from A, so that $k_{AC} = k_{AB}k_{BC}$. Accordingly we find that k factors simply multiply. If you express the k factors in terms of the relative velocities v_{AB}, v_{BC}, v_{AC} by means of equation (1) you obtain Einstein's famous law of the composition of velocities:

$$v_{AC} = \frac{v_{AB} + v_{BC}}{1 + v_{AB}v_{BC}}. \tag{2}$$

This result was in part discovered by Fizeau long before relativity when he measured the velocity of light in fluids moving fast along pipes. Thus here we have a test of the theory available from work antedating the theory by many years.

The results so far obtained give an impression of the power of the k-calculus. For our own thinking the most significant conclusion is that we must abandon the elementary concept of one universal time. We must return to the basic idea that any quantity in physics is defined by the method of measuring it. Therefore time is that which is measured by a clock. There is no reason to believe that all clocks, irrespective of their states of motion, should read the same time. Indeed, *my* time is what I measure by *my* clock, and another observer, differently moving, measures *his* time with *his* clock. If both he and I move inertially, then by the principle of relativity we are equivalent. His time is as good for his circumstances as my time is for mine. This discrepancy of times is very clearly displayed in the work leading up to equation (1) (see also Fig. 4), where B measures with his watch time kT between his receipt of A's two light signals, while A, using his own clock, assigns a time interval $\frac{1}{2}(k^2+1)T$ to the separation of these two events.

How, then, do we come to have an intuitive idea that all time is the same, an idea we now regard as an intuitive mistake? Equation (2) shows this very clearly. The intuitively obvious

$$v_{AC} = v_{AB} + v_{BC} \tag{2'}$$

results from (2), provided the product $v_{AB}v_{BC}$ is negligibly small compared with unity. Remembering that our quantities v are the ratios of the velocity of our objects to the velocity of light, we note that in daily life all v quantities are very small indeed. Even the speed of a jet air liner is barely 10^{-6} in our units and thus we can work with (2′) instead of (2) unless we experiment with high-speed particles or use extreme accuracy. Equally, by the converse of (1), 'daily life' values of k differ from 1 only by minute quantities, and hence kT and $\frac{1}{2}(k^2+1)T$ differ by less than one part in 10^{12} in these circumstances. The experience on which our intuition is based is too restricted to have revealed to us the multiplicity of measured time. We can learn about this only from experiments dealing with high velocities. Then indeed we find that we must use (2) rather than (2′) and must be guided by experiment rather than by our inadequately grounded intuition. Thus relativity is appreciated as the physics of velocities higher than those we ordinarily meet.

I now want to discuss the infamous clock paradox (so-called) because it leads on to some of the other things I want to say. First I want to draw some simple conclusions from the results we have just reached. Consider Fig. 5, and suppose that observer C is at rest relative to observer A. Then $k_{AC} = 1$. This follows directly from (1), since $v = 0$ (relative rest) implies $k = 0$. Thus in this case k_{BC} is the reciprocal of k_{AB}. Also from (2), since $v_{AC} = 0$, $v_{BC} = -v_{AB}$, which means simply that B approaches C as fast as it is receding from A. Hence the same velocity that gives a factor k in recession gives a factor $1/k$ in approach.

Now we can readily examine the so-called clock paradox. We again take three collinear observers, A, B and C (Fig. 6), the velocity of B and C relative to A now being the same, but in opposite directions. Accordingly, if factor k applies between A and B, factor $1/k$ applies between A and C. The timings are so arranged that first there is the encounter of A and B, then that of B and C, and finally that of C and A. By the very symmetry of the figure, B's measurement of the time interval T between his two encounters (with A and C respectively) must equal C's measurement of the time interval between his two encounters (with B and A respectively) which therefore also equals T. If B and C each emit light signals at each counter, then A measures interval kT be-

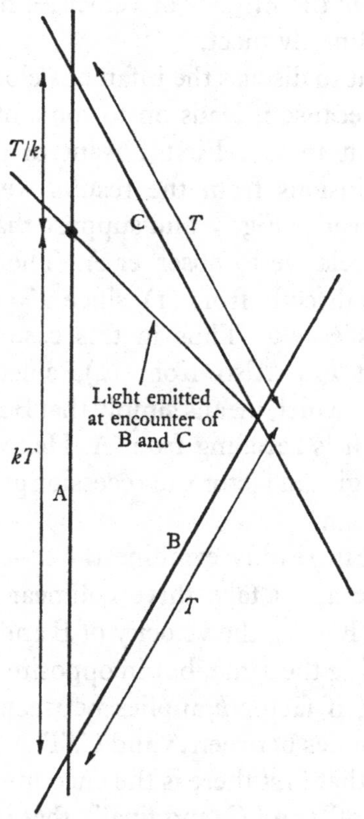

Fig. 6. The 'clock paradox'

tween the receipt of B's two signals, and T/k between the receipt of C's two signals. B's first signal arrives instantly at A's first encounter, C's first signal arrives together with B's second signal, while C's second signal is immediately received at A's second encounter. Hence A measures a lapse of time

$$(k+1/k)T = (k^2+1)T/k$$

between the first and the last encounter, while the combined measurement of B and C equals $2T$, which is always less than A's measure by a simple inequality.

This result bothers some people because they say: 'why this asymmetry of measurement in what appears to be a symmetrical situation? You happened to have chosen A to be at rest in your system of reference, and B and C moving. Surely we could have done the opposite, taking B and C to be at rest and A moving, and then the inequality should have been the other way round, the inequality being the fact that A's measurement of time is greater than the combined measurement of B and C?' But this is sheer nonsense. Of course I could have carried out the same calculation taking B to be at rest in my frame of reference and A and C moving, or taking C to be at rest in my frame of reference and taking A and B moving. I would have got exactly the same result, which after all only depends on the factor k, which does not refer to any frame of reference. (The concept of frame of reference is irrelevant to my calculation, but only applies to the purely indicative drawings of Fig. 6.) The one thing I could not possibly do is to take B *and* C to be at rest in any frame of reference because B and C have a relative velocity.

So one time interval (the shorter one) results always from the combined measurement of *two* inertial observers, whereas the other is the measurement of *one* observer. There is hence no symmetry between the measurements; relativity never says that $2 = 1$.

Now you may say: all right, but instead of working with the clocks of the two observers B and C, let us ask B to throw his watch to C, who neatly catches it. Now there is *one* watch measuring on each side. Why should one of these watches give a longer time than the other, since they are equivalent? But this argument is also wrong. The two watches are in no way equivalent. A's watch leads a perfectly good existence as an inertial watch all the time. The other watch suffers an acceleration, a change of velocity, during its transfer from B to C (and is thus non-inertial). The sharper I make the curve (Fig. 7), the shorter the time in which the watch has to change from B's velocity to C's velocity, the greater its acceleration. There is nothing in the principle of relativity that says a non-inertial motion is equivalent to an inertial motion. On the contrary, the very act of singling out inertial observers implies the inequivalence of non-inertial ones. Moreover, we know from experience that clocks react in particular ways to non-inertial motions. A well-known non-inertial motion to which one may subject one's watch is to drop it on the floor. In certain cases this has quite profound effects on its time-keeping, effects that are never found with watches moving inertially. There is then a clear-cut difference in that one watch has a non-inertial motion where A's watch has an inertial motion. Accordingly

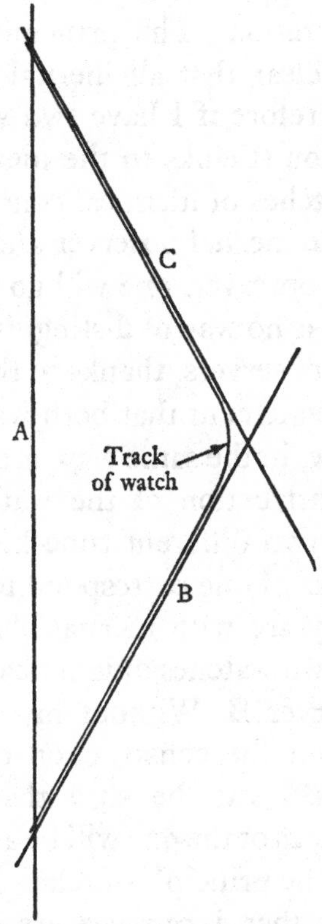

Fig. 7. The 'clock paradox' with non-inertial motion of watch

they are not equivalent, and it is perfectly reasonable to say that a watch following on inertial motion indicates a longer time between two events than a watch in non-inertial motion. There is nothing self-contradictory here at all.

But there is a point of further significance here. What precisely happens to the watch when I subject it to this acceleration? The principle of relativity makes it quite clear that all inertial observers are equivalent. Therefore if I have two watches of the same construction (thanks to the identity of atoms we can have watches of identical construction), and I put one on one inertial observer and the other on another inertial observer, one will go just as well as the other; there is no way of distinguishing between the two inertial observers, thanks to the principle of relativity. The statement that both watches will go, in a certain sense, in the same way, is quite independent of the construction of the watches. Indeed, suppose I take two different time-keeping devices such that 98 ticks of one correspond to 1 tick of the other when they are with inertial observer A, and suppose I give two watches of identical construction to inertial observer B. Without my knowing anything at all about the construction of these time-keeping devices I can be sure that, on inertial observer B, 98 ticks of the one will be as long as 1 tick of the other, by the principle of relativity. The time-keeping devices that I can use are complete unrestricted. As I sometimes say, you can take atomic clocks or you can count the frequency with which you felt hungry; or you can count generations of rabbits, or use beta-decay. Any of these things work.

The only type of time-keeping device that I would like to warn you against is the pendulum clock; and the reason why one has to be careful with this device was made clear to me at a conference ten years ago. A number of us were puzzled about this matter, and then, in one of his last contributions, the great physicist von Laue said: it is quite simple; a pendulum clock is not the box you buy in a shop; a pendulum clock is the box you buy in a shop together with the Earth. If you want to transfer a pendulum clock from one observer to another you have got to supply them both with Earths, which is a little on the expensive side!

But if we ask how the clock we are considering reacts to the period of acceleration, then this important question cannot be answered without knowing something about the construction of the clock. This is a very fundamental point. All we have said before about the equivalence of clocks on different inertial observers were completely general statements applying to all kinds of time-keeping devices. But how a clock reacts to acceleration is utterly dependent on how the clock is constructed. If, for example, our clock is a rabbit generation counter, and the acceleration is so high that all the rabbits die, then this particular device won't work any more. Similarly we know from everyday life that there are shock-proof watches and watches that are not shock-proof, and that these respond rather differently to the same acceleration. So whether our clock can survive or not is a question of our choice of clock. We know that there are clocks that can take very high accelerations: for example, radio-active nuclei. But they cannot survive enormous accelerations, i.e. indefinitely large ones, either. If you want to give a nucleus a very high acceleration, you can only do so by bombarding it with others, and

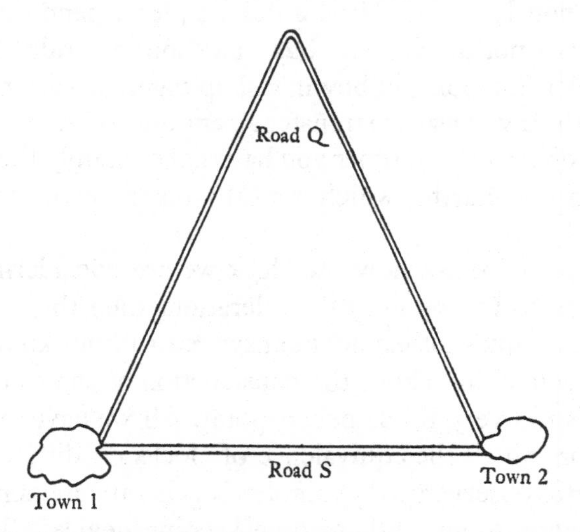

Fig. 8. This sketch shows that road Q is far longer than road S though road Q is straight nearly all the way

if you hit it too hard it will bust—just like the watch. So, every kind of time-keeping device, from the rabbits to the radio-active nucleus, has a limiting acceleration which will bust it; and somewhere, a little below it (again depending on the construction of the clock), a limit above which it won't work properly. The moment I use the expression 'won't work properly' I am in rather difficult country, because how do I decide whether a watch works properly or not? We know from the principle of relativity that all watches work equally well on inertial observers. I can therefore say that whenever I have a curved track in my diagram—which means an acceleration—then there will at any moment be a tangent inertial observer that is an unaccelerated observer moving momentarily at the same speed. I can now define a clock working properly as one that, during the time it moves together with this inertial observer, works just like a clock of the same construction on the inertial observer. This is the usual geometrical way of rectification where we define the length of a curve as the sum of the lengths of the straight lines by which it can be approximated. We use just the same definition here for time measurement.

Some people have argued rather foolishly: how can it matter for the time difference between the inertial and the other clock that this second one is accelerated, because I can make the period of acceleration very short? This is just as sound as to draw lines on a map between two towns (Fig. 8) and to say: road S is the shortest road between the two towns because it is straight; road Q should, by the same argument, be of virtually the same length because it is straight nearly all the way. You see the deep truth in this? Road Q is long, not because of the length *in the curve*, but because it is *curved*. And similarly the accelerated clock gives a time shorter than the inertial one, not because of the length of time it is accelerated, but because it has a period of acceleration on the track.

But I want to stress again the point I have already made—that having a period of acceleration means that I have to have some technical information about my clock. I can tell you how my clocks behave on inertial observers without knowing how they work. If I want to know how they behave during a period of acceleration, then I have to know how the clock works. Different kinds of clocks will always keep in step in the same way on all inertial observers; they may well behave differently when periods of accleration are involved. And this is just once again a way of stating that there is an absolute difference between being inertial and not being inertial. There is a true equivalence between all inertial observers which does not extend to non-inertial observers. This is the essence of the principle of relativity and of the whole theory.

3
GRAVITATION

The subject discussed in the last chapter is called the special or restricted theory of relativity. The reason for the restriction or the speciality or the nature of the limits has frequently been misunderstood. The restriction lies in confining the validity of special relativity to the case in which there is no gravitation. It has nothing to do with whether I do or do not discuss accelerated motion. The statements that special relativity only applies and that I can use the theory only for inertial motion—which are unhappily made quite frequently—are completely incorrect. One can work out what happens in accelerated motion just as easily. Indeed I gave you an example when I showed how one can compute the time along a curved track by approximating it by a succession of inertial observers. Equally one can consider the stresses induced by the acceleration, very much as in Newtonian theory. This shows how the apparatus of the theory can be used to calculate what happens to a non-inertial observer. Of course, as I have stressed several times, the experiences of the non-inertial observer will not be the same as those of the inertial ones. But I can use special relativity to calculate the experiences of the one just as much as those of the other.

We can come to the next stage in the development, the theory of gravitation, in several different ways. First let me give you the historical way in which Einstein approached it. Einstein's desire was to show that equivalence was not wholly restricted to inertial observers, but that there was also equivalence extending to accelerated observers. You will rightly ask, when I have stressed so much the enormous difference between an accelerated and an unaccelerated observer, how Einstein could hope to find any kind of equivalence between the two. After all, when I am accelerated I do suffer all sorts of troubles like the watch that fell on the floor, or the rabbits that died because of the high acceleration. But Einstein remembered another very important result in physics, a result due to Galileo, that all bodies fall equally fast. You appreciate that when I move in a non-inertial way the bodies that are with me will seem to try to get away from me. Best known of these effects is the apparent centrifugal force which I observe when I am rotating and am unaware of the fact. The Coriolis force and the ordinary linear acceleration 'forces' are all very well known. All these forces, brought into being by my own non-inertial motion, have one feature in common: What is prescribed, as it were, is the acceleration. If I have a certain acceleration, then I see other bodies moving with minus that acceleration relative to me. Therefore I ascribe to them forces *proportional to their masses*, by Newton's second law. Thus these fictitious forces that appear to be acting on bodies solely due to my choosing to live in a non-inertial frame are all proportional to the mass of the body they are acting on.

But, as Galileo pointed out, there exists another force that has the same characteristic, namely gravitation. It also has the characteristic that it gives the *same* acceleration to all bodies, or, if we use Newton's second law, the force is proportional to the mass of the body. Hence gravitational forces and the fictitious forces due to the observer's non-inertial motion are of the same kind as far as the reaction of various bodies to them is concerned.

Both these forces affect all bodies equally and give them all the same acceleration, while all other forces differentiate between bodies. For example, an electrical field will exert a strong force on a charged body, and leave an uncharged one quite unaffected; a magnet will attract iron but not glass, etc. However, gravitational forces and the fictitious forces due to my own acceleration cannot be distinguished by any comparison between different bodies such as charged and uncharged ones. When I am driven in a car on a bumpy road full of potholes my obvious reaction is to say: I am getting bounced around because this car is following a markedly non-inertial path. But I could equally well say: this car is following a perfectly inertial path but we have a strongly time-variable gravitational field today. I must confess that on some days when I am a bit clumsy and drop things right and left, I feel not only that on that day gravitation is particularly strong but that it has a powerful time-dependent horizontal component. You could interpret your experiences in the car on the bad road in just the same way. You do not choose to do so because you have acquired certain prejudices about the constancy of gravitation, you have acquired a certain knowledge of what the road appears like when you look out of the window, and so on. But if you did not have these prejudices and did not look out of the car, then you could not tell from observations within the car which was the case: was the road bad or was gravitation just playing tricks?

In this sense then we can speak of a form of equivalence between the non-inertial observer and an inertial one. I personally like to look at it a little differently. I should like to remind you of the fact that in our development of special relativity we stood four-square on Newton's first law of motion. And this is terribly important. It was this that supplied us with the concept of the inertial observer, with the whole set of inertial observers having a relative unaccelerated motion, and so on.

Now if we have such an important law then it is not unreasonable to ask for it to be tested. Imagine then that you are asked to test Newton's first law. Clearly, what you will have to do is to find a body on which no forces act. So you choose a body and then you make sure that no forces act on it. You might fear that somebody is pulling strings, and so you go around that body and cut every string you can find. You might fear that it is being pulled by magnetic fields, so you measure whether the body has any magnetism and, if it has, you neutralize it. Similarly with electric charge. And in this way you can abolish one force after the other, until you come to gravitation. By Galileo's result, there is no cure for gravitation. You will therefore say, in a very despondent manner: awfully sorry, I can't test this law here because I can't abolish gravitation. Newton's answer was: go elsewhere; find some place where there is no gravitational field. This is not just expensive, but genuinely rather difficult. The universe seems to be full of matter, and you don't seem to be able to get arbitrarily far from all matter. Anyway, we want Newton's first law *here* and not out there, because we use it for our physics here. And so we have the very awkward situation that Newton's first law—on which the whole concept of inertial observers is based and on which all relativity is based—is untestable. And an untestable law is not a scientific one.

What does one do in this situation? What Einstein did was to say: very well, if we can't abolish gravitation, let us say that gravitation and inertia are the same; after all, they attack a body the same way; the forces resulting from both are proportional to the body's mass. Accordingly we re-word Newton's first law, dividing it into two parts. Part I of Newton's first law now reads: there exists a standard motion of bodies such that, if bodies follow that standard, no force is acting on them. And part II states that this standard motion is motion in a straight line with constant velocity. Einstein proposed to drop part II. We still have a standard motion, but this is a motion following inertia *and* gravitation. Only if a body deviates from such a motion of free fall do we speak of a force; and such a force is then something which pulls things off their standard tracks and therefore is not gravitation. Gravitation is not a force; gravitation is one with inertia in determining this background of standard motions. Of course these standard motions are complicated; the totality of motions of free fall are a great deal more complicated than motions along straight lines with constant velocity. How much more complicated? If I just think of something the size of a room, then in the room there is a constant gravitational field g, which is the common acceleration of all bodies, and their motion will be unaccelerated in free fall apart from this common quantity g. So if I subtract out this g to obtain the relative motion, then the relative motion of all bodies falling in the room will be unaccelerated. It is clear that this is only because I have regarded g as constant within the room. If I want to extend the area of my observations to cover, shall we say, the whole Atlantic, then it is obvious that g on the American coast will be inclined at about 50° to g in England. Therefore, when viewed from objects falling freely in England, objects falling freely in America will not look as though they were unaccelerated.

Let me put it a little differently. If you fall freely

yourself, you abolish gravitation in your vicinity. (This is the weightlessness in a space ship.) What you do not abolish, however, is the *non-uniformity* of gravitation, which leads to a relative acceleration of neighbouring particles. This relative acceleration of neighbouring particles is important. In our journey round the sun opposite sides of the Earth can be considered neighbouring particles, and the gravitational field of the sun is not the same on these opposite sides. This is the cause of the solar tides on the Earth. The same effect with the moon gives us the lunar tides. So the unabolishable part of the gravitational field is the relative acceleration of neighbouring particles. If you then want to make a theory of gravitation, the essential intrinsic effect you must describe is the relative acceleration of neighbouring particles. You must, however, take account of special relativity. As we have seen, special relativity is founded on Newton's first law. This law does not apply in a gravitational field in the large. Because of the non-uniformity of the field, freely falling observers far from each other have a relative acceleration. But it does hold in the small, if we refer one freely falling particle to another and neglect their relative acceleration. How small the region of validity is depends on the accuracy with which you measure things. Specify an accuracy and I can describe a region small enough for the relative acceleration of neighbouring inertial observers to be irrelevant, and then Newton's first law—and therefore special relativity—will apply to freely falling bodies in the region. Thus we want a theory that describes the relative acceleration of neighbouring bodies and that, when we neglect this, reduces to special relativity. This theory is Einstein's general theory of relativity. It happens to be mathematically rather complicated; there are good reasons why it has got to be complicated. It is therefore rather an awkward theory to handle. But we have got a great deal of confidence in it. First of all, because it is relativistic—that is to say, it reduces to special relativity in the small. We have countless observations that test special relativity. Secondly, on the large scale, when we have weak gravitational fields, it reduces to Newton's theory of gravitation, which has been tested so thoroughly in the solar system. Finally, there are certain minor deviations from Newton's theory even in the solar system and, as far as our observations go (these effects are rather small), the tests go rather well for general relativity.

The only point I want to make here is that, being a relativistic theory, it is immediately capable of handling the fastest of all things, that is light; whereas Newton's theory, being a purely dynamical theory, didn't, without additional assumptions, tell us anything about light at all. It is particularly in treating the interaction of gravitation and light that general relativity is essential to us.

In the last part of this chapter I want to refer to certain recent discoveries in general relativity—purely mathematical, theoretical discoveries—which

I at least find extremely disconcerting. First I will explain why I feel they are disconcerting, and then I will tell you what they are.

When I suggested that the man in the bumping car might think that the gravitational field was highly time-dependent that day, one gave vent to one's prejudices about gravitation and regarded this view as absurd. These prejudices are: (1) appreciable amounts of gravitation are only produced by very large sources; (2) these large sources are very noticeable to us by other means; (3) these sources have a great deal of permanence. As for (2), we don't only know through the existence of the Earth's gravitational field that there is an Earth under us. We know about the sun other than from the Earth's orbit about it. The mass, which is the source of the gravitational field, satisfies a well-known conservation law, and the momentum (which then effectively describes the velocity with which the centre of mass moves) satisfies another conservation law. Therefore gravitation has to be very well behaved, its sources have a certain amount of permanency about them, they are very noticeable by means other than investigation of the gravitational field, and appreciable amounts are only generated by rather large masses. I would like you to remember that, if Newton had accounted for the Earth's motion by referring to a source of the field not noticeable by other means, we might still have swallowed the theory, but not nearly so readily. It was because it was a well-known object, the sun, that was found to be responsible, that the theory was so easily accepted.

I have mentioned the fact that general relativity differs from Newton's gravitational theory in practice only in some minor observational matters. In what imaginable circumstances does the theory differ markedly from Newtonian theory? There are basically two such circumstances: (1) when the gravitational potential is large. By this I mean that a body has a certain amount of gravitational potential energy, which is a multiple of its mass. By special relativity it has a certain energy corresponding to its mass which equally is a multiple of its mass. If the first of these energies, the gravitational potential, is a sizable fraction of the energy associated with the rest mass of the body, then we have a situation in which we get appreciable differences between Newton and Einstein. Note that it is the potential that matters, not the field. We can get very large potentials with modest fields if they extend over very large regions, and this is a point worth making, because it is often forgotten. (2) When fast motions are involved, general relativity is applicable because it is a relativistic theory. It also means that everything in gravitation is subject to the same rules and regulations as everything else in special relativity, one of the rules and regulations of which is that no information can travel faster than light. We can then ask what happens if the sources of the gravitational field move fast, and by fast motion (of course, velocity itself does not mean anything) I mean either two sources in high relative velocity, or one source that in some way changes its field rapidly.

Now remember that, in a somewhat more intricate way than in Newtonian theory, there are also conservation laws for mass and momentum in general relativity. The old question of how soon would the Earth notice in its orbit if the sun ceased to exist, is not a sensible question in general relativity because there is a conservation law for the mass of the sun—the sun just can't suddenly cease to be. But it is equally silly to ask what would happen if the sun suddenly walked off at right-angles to the Earth's orbit, because there is a conservation law for momentum and the sun just can't do that. But what can happen to the sun is that it can grow horns, and there is no conservation law that stops that, only the sun's inherently good character. In other words, the sun may change shape from a sphere to a prolate spheroid. No rule stops it doing that at any speed it likes, as long as no particle there has to move faster than light. A spheroid has a different attraction from a sphere, as follows from Newtonian theory.

We can therefore sensibly ask the theory to tell us about the behaviour of the gravitational field due to sources that are up to tricks of this kind, and the theory must give an answer. The theory very decently does give an answer. It tells us that the Earth will leave its orbit at just the moment when we can *see* the sun changing shape. Gravitational information travels with the speed of light. This is one problem out of the way in just the manner appropriate to a relativistic theory.

Problem two: anybody who has sent a telegram knows that it costs money to send information; it is a general rule in physics that you cannot convey information without energy. If the sun changes to a spheroid, and, through the consequent change of its attraction, spreads information about its change of shape, then it must lose energy, and hence mass, and so attractive power. And again the theory gives a very sensible and reasonable answer to this loss of mass due to the sending out of gravitational information.

The subject I am writing about, incidentally, is called the theory of gravitational waves. Unfortunately these information-conveying energy-carrying gravitational waves do not travel quite like light. The difficulty here is this (I become frightfully mathematical for a moment—so mathematical that I could not possibly write down anything). When I have a second-order partial differential equation classified as being hyperbolic, then it means that broadly speaking the solutions have wave character. That is to say, there is an arbitrary function of the time that I can feed in (corresponding to the shape of the sun in the gravitational case, corresponding to the current in the transmitting aerial in the radio case, etc.). The information from the transmitter then travels at a speed not exceeding the fundamental velocity of the equation—the speed of light, in our cases. But there are two quite different possibilities that you can distinguish mathematically. In case one *all* the perturbation travels at the speed of light. In case two some travels at the speed of light and some lags behind. This lagging behind is called the tail of the wave. If we consider the ordinary wave equation—say the equation of sound in three dimensions—then it is of type one, and all the information travels at the speed of sound. If I make a sudden noise now, then the noise will be always confined to the surface of an expanding sphere. None of it will be left behind inside that sphere. However, if I arrange my sources on a straight line and bang them all simultaneously (for example, if I explode a lot of fireworks along a straight line), then, as you can imagine, I get a cylindrical wave travelling out, not a spherical one as from a point. When you work it out then this is of type two. The bang is not confined just to the surface of the cylinder but some of the excitation is left behind. Of course you can tell yourself that there is a simple explanation, because at any point I get information about the bang not only from the nearest point of the line (from where it reaches me first) but also from points further away from where it will reach me later, resulting in a tail of the wave. This particular explanation is unsatisfactory because, if instead of having a line of charges I have a plane of charges, then the propagation is again of type one, as in the spherical case. All the excitation is in the wave front and nothing lags behind.

Gravitational waves have, so it seems, tails, i.e. are of type two (we are not quite sure; mathematically it is very complicated, but we are pretty certain). This is the only way in which we can possibly understand a relation found by Newman and Penrose recently; it is a strict conservation law. I mean by 'strict' that the quantity concerned cannot change at all. (By 'non-strict' I mean cases like the conservation of mass, where, as I have mentioned, if the sun radiates a lot gravitationally it loses mass.) There is a quantity—to be precise, in the general case, there are 10 quantities but in the axially symmetric case one quantity—that is strictly conserved whether gravitational waves are sent out or not. It is a quantity of the second order that is evaluated asymptotically, i.e. far from the source. Newtonianly speaking (and I may speak so whenever the system is almost static) it involves the products of mass and quadrupole moments and squares of dipole moments and that sort of thing. This result I regard as horribly uncomfortable. Suppose I have a body that is in the shape of a static spheroid; then I can work out its gravitational field completely and exactly in general relativity—the variety of fields turns out to be just as rich as in Newtonian theory. Then this quantity, this conserved quantity—I am talking now of the single one because of the axial symmetry—has a certain value that is not zero for such a spheroid. If we next consider a sphere, then again we can work out this quantity, which now turns out to be zero.

Now let us suppose that the spheroid, after having been a spheroid for a semi-infinite time, gets tired of its shape and changes into a sphere. In the transition it will of course send out gravitational waves, but after that it will settle down happily and live as a good sphere for ever after. But its field can never be that of a sphere. Its past is pursuing it, because this quantity that I have been speaking about is absolutely conserved and therefore cannot change its value from non-zero (for the original spheroid) to zero, corresponding to the static field of a static sphere. So however long I wait thereafter I can never say that the field of the body is exactly the field of a sphere. There is nothing in principle against this; it means that a gravitational field depends not only on the state of the source at the time I see it, but also on its past history. What is so serious and awkward is that the field *never* forgets; the quantity in question is *strictly* conserved, and so all the history down to minus infinity matters. This, it seems to me, destroys a great deal of the appeal of what Newton said: that the source of the gravitational field is manifestly apparent, that you can see it. If I investigate the field in detail it isn't just what I see *now* that matters, but all that has happened to the source from the year dot onwards. I presume that it is some of the tails of these waves that haven't got right away and that have stayed behind that are responsible for this, because the quantity concerned, as I explained, has to be evaluated asymptotically at infinity. Of course, I may perhaps have been a little remiss in calling this body a sphere, because, if I assume from the way things are at infinity that space itself isn't perfectly spherical, then the body can't be a sphere in a space that is a little uneven. It just tries to be as good a sphere as it is allowed to be, only it is not quite perfect.

I personally think that this was one of those disastrous discoveries—like, say, the discovery of the quadrupole moment of the deuteron—which set back our thinking by many years. But occasionally such discoveries come along. This result of Newman and Penrose suggests that the connection between the field and its sources (as observed by other means) is unhappily not quite as simple as one would wish it to be. If there had been an exponential decay down to the spherical value I would not have minded a bit—I've got patience and can wait. But with the quantities never changing, I am baulked, and have to ascribe some of the properties of the gravitational field to the dim and distant past.

4

THE ORIGIN OF INERTIA AND THE UNIVERSE

In the previous chapter I spoke about the way we are driven, through the impossibility of testing Newton's first law, to regard inertia and gravitation as one, and about how this forms the basis of general relativity. We found that inertia and gravitation had to be considered as one and the same thing in their *effects* on bodies. But what about their sources? The source of the gravitational field is clear enough, or so it seems. What about the source of inertia? This brings us to the famous question known as the problem of Mach's principle. Roughly speaking, it is that in Newton's theory certain states of motion are selected as being called inertial. What defines this selection? What determines them? In Newton's view you took an absolute space, and inertia was then counted relative to this absolute space, so that acceleration was an absolute, as I have previously pointed out.

There are several very grave disadvantages in this. First, we can quite sensibly ask if there is such a space why is it so well hidden? Why doesn't it reveal itself through velocity at least, if not through position? Why do we have to go to the second derivative, to acceleration, before we find out anything about it? Much more serious is the criticism that the word 'absolute' space doesn't explain anything. It is not a hypothesis that can be tested in any way. It is just another way of saying something we know already, namely that acceleration in the Newtonian scheme of things is absolute; a repellent idea to a modern physicist. The opposite attitude, Mach's principle, was perhaps put most beautifully by Einstein himself when he said that in a consequential theory of relativity there can be no inertia of matter against space, only an inertia of matter against matter.

With this formulation Einstein clearly identified the sources of the inertial field as being material. However, there are grave difficulties in identifying and finding these sources. Normally when we work in physics we work in our laboratories with things that can be switched on and off, and then one can easily establish a causal link—when the switch is on, A happens, when the switch is not on, A does not happen. Whatever the switching on does, it may be presumed to have some reasonably close linkage with phenomenon A. We were much concerned about the difficulties of switching off gravitation, and these are real enough and cause worry enough. But nevertheless we do know a great deal about different gravitational fields throughout the solar system, and the basic rules of orbiting were found by Kepler a good many years ago. I sometimes feel it would be nice to speculate how physics would have developed if the solar system had consisted only of the Earth and the sun. Conditions for life would have been indistinguishable from the actual, but the whole development of many important branches of science would have been quite different. People's interest was focused on astronomy by the motion of the planets, and the work on this subject culminated in the researches of Kepler, Galileo and Newton. If we had not been presented with this magnificent outdoor laboratory with so many different things happening, if all we had known was that the Earth was following an orbit round the sun, it would have been far more difficult to progress, quite apart from the fact that early interest in astronomy would not have been aroused.

But to come back to my basic point, although we cannot switch gravitational forces off and on, we yet see plenty of examples in the sky of bodies moving under different gravitational fields. But with inertia the problem is clearly very much more difficult. The fact that Newton's theory describes the motion of the planets and satellites so very closely proves that the inertial frame is effectively one rigid frame throughout the solar system. In other words, the inertia-causing effect of the bodies in the solar system—the sun and the Earth and Jupiter and the moon—must be completely negligible. If we regard these bodies—if we regard matter—as the source of the inertial field, then at first sight the evidence is all against it because all the big bodies we know seem to have negligible effects on inertia.

How can we nevertheless save the idea? Only if we

appeal to a higher court than our astronomical neighbourhood; if we go further afield. If the sources of the inertial field are overwhelmingly at large distances and not the bodies nearby, then it would not be unreasonable to have an inertial field that was virtually constant throughout the solar system. We cannot readily assume a law of force where the force is greater the further you are from the source. This would go very much against the grain. But nothing of that nature is necessary. If we have a law of force varying, say, inversely with the distance, then certainly the very distant bodies would win hands down over the near ones because their total mass is so very much greater. With an inverse square law the effects of near and distant bodies are neatly balanced (Olbers' paradox); with an inverse fourth power law the near bodies are vastly more important than distant bodies. This leads us to the basic idea that if the sources of the inertial field are identifiable as matter, then the law that determines how much inertia is due to a body must not depend too strongly, through too high a power, on the distance from that body.

I have perhaps spoken a little glibly about identifying the different sources of inertia, because that is a terribly linear way of thinking. If the inertial field is a highly non-linear field (which is entirely possible), then it is unreasonable to ask to pick out individual sources. On the other hand, it leads us to the conclusion that inertia, if we can account for it in any rational way at all, must be closely linked with cosmology. There is nothing else we can link it with. This is very sad—it is much better to try to link phenomena with things you know something about rather than with things you know next to nothing about. But since we can exclude all other culprits, we must link inertia with the structure of the universe as a whole.

If we had a theory of Mach's principle, how would we test it? If such a theory is more than a parlour-game, if it pretends to be scientific, then it must be testable by empirical means. It does seem to me that this is a really great difficulty. There are various ways in which one could conceivably have a Machian theory that, at least in principle, is testable. Even if the effect of the very distant masses is overwhelming and far greater than that of the sun, nevertheless there might be a residual slight effect of the sun which you might be able to detect. Again, the universe is not all that homogeneous. We are fortunate for these purposes in being in a very eccentric position in our own galaxy. On the inverse distance law, our own galaxy would bear more responsibility for inertia in our area than the sun or the Earth, and we might perhaps have an effect due to our own galaxy. It could be, as some authors think, that mass itself, inertial mass, ought not to be a scalar but a tensor quantity, so that inertia varies with direction. And it is not inconceivable that in some such way we could test the theory of Mach's principle.

This is the real excuse for going on with it. It does not mean either that we have a theory or that the test is round the corner. But it is necessary before attempting to think very seriously about Mach's principle to show that it is not something that is necessarily untestable.

The first person to take Mach's principle seriously was Einstein, and by taking seriously I mean that he really tried to build his theory round it. What he did in fact was that, a year after the formulation of general relativity, he introduced a modification into the theory in the hope that thereby it could satisfy Mach's principle. The test that he applied, the theoretical test, was the following. In relativity, given the sources, you have equations for the total field, that is for the combined inertial and gravitational field determining the standard motions, i.e. the paths of free fall. What Einstein found was that if you took unmodified general relativity and assumed that there were no sources whatever, you nevertheless got a solution for the combined field. This solution turned out to give a field without gravitation so that everything moved nicely in straight lines with constant velocity. He considered this result to be quite unacceptable. If inertia is due to matter, then in the absence of matter there should be no inertia. The motion of an isolated particle should thus be completely undetermined. The aim of the modification that he introduced was to see to it that the equations had *no solution* in the absence of matter, while giving reasonable solutions when matter was present. He tested that the modified equations had a reasonable solution in the presence of matter and thought by a plausible mathematical inference that they were indeed without solution in the absence of matter. However, this inference turned out to be incorrect, as de Sitter showed shortly afterwards that these modified equations had a mathematically admissible solution in the absence of matter. This solution, though originally regarded as only a mathematical curiosity, was in fact the first solution for an expanding universe.

Einstein said: 'If my modification does not work I will throw my modification away' and from then on worked only with the original unmodified theory. However, not long afterwards the astronomers discovered the red shifts in the spectra of the galaxies. These were interpreted as fitting into the expanding type of universe first found by de Sitter. Accordingly, in spite of Einstein's own rejection of the modification, it has been of much importance in theoretical cosmology.

I don't want to go into the later history of this so-called cosmological term, but I do want to say that in my view the test that Einstein applied was inappropriate. I take an extremely empirical view of general relativity. It is a theory like all other physical theories, founded on experiment and observation. We expect a decent theory to go a *little further* than the experiments and observations on which it is based, to account for something a little more general than the particular circumstances in which the theory became established. But we never expect our theories to hold in circumstances utterly and completely different. All our observations and experiments have been carried out in the presence of a material universe. To expect that a theory so based should work in the absence of a material universe seems to me an entirely unjustified step. Our theories, even the best ones, are nowhere near good enough to take an extrapolation of this magnitude. One could set out a little history of knowledge which showed how every time experimental accuracy increased by a very few powers of ten, entirely new and previously undreamed of concepts had to come in. To suggest that we can drop the density of the universe by an infinite number of powers of ten without bringing in completely new concepts does not seem to me to be at all plausible.

Nowadays one can say that there are three separate schools of thought about the relation of Mach's principle to general relativity. There are some people who regard the whole of Mach's principle as a pseudo-question, unimportant because it is so far from leading to a testable theory. There is a second school of thought which says that general relativity is all right and does not need to be modified, but it must be embedded in the appropriate universe. We would not expect a theory derived from local gravitational observations and other observations and experiments (those relating to special relativity which is so basic to it) to be viable in a general model of the universe quite different from the one we are in. As soon as you put general relativity into a model universe roughly like the one we have got, Mach's principle will emerge, according to this view. Here of course progress is somewhat held up because of our grave uncertainties about the structure of the universe. But we can say, with our present-day knowledge, that a multitude of reasonably plausible models of the universe, combined with general relativity in some way or other, yield acceptable results in terms of Mach's principle, in the sense that the local inertial frame seems nicely tied to the distant matter in the universe. The third attitude is that this is not good enough, and you must have a theory that really accounts for Mach's principle in some detail. There are some quite powerful arguments in this direction. The one that impresses me most is the one about the constant of gravitation. In as far as we can give any real meaning to the constant of gravitation, it does express the relation between the gravitational and the inertial properties of matter. The gravitational ones, we have every reason to believe, are directly connected with the local sources. As I have been saying, on any theory of Mach's principle the inertial properties are connected with the distant sources. If we have an evolving universe—which of course we might not have—then the structure and lay-out of the distant sources will be changing in time. One must then contemplate, to say the least, that the constant of gravitation itself will change in the course of time, because the inertial properties of matter determined by the distant lay-out of the universe may then be changing. Unless you have a theory of just how these inertial properties are linked with the distant matter, and how they depend on the distant regions, you can't work this out. In general relativity—at least as ordinarily interpreted—we do seem to be making the assumption that the constant of gravitation is a true constant. So here one can see a real argument for making a theory of Mach's principle. Doing something about it, of course, suffers from the great difficulty, not only that observational tests seem to lie far in the future, but (if you follow me in my attitude to the empty universe) theoretical tests are not much use either. So we are left with a wide open question.

Various attempts have been made to deal with this problem, and I want to refer quite briefly to the recent attempt by Hoyle and Narlikar.[1] I find this particularly interesting because it brings in another old hobby-horse of mine, the relation between particles and fields. One can see from the example of Newtonian theory how this arises. I can write down Newtonian gravitational theory in two different but almost equivalent ways. I can either write down that the Laplacian of the gravitational potential is given by the density, multiplied by a suitable factor, or I can say that my gravitational potential is proportional to the sum $\Sigma(M/R)$ where the mass M of each particle of the universe is divided by my distance R

[1] Hoyle, F., and Narlikar, J. V., 'A New Theory of Gravitation', *Proc. Roy. Soc.* A, **282** (1964), 191–207.

from it. These are two equivalent formulations. A well-known exercise proves that they are equivalent.

But in fact this is not strictly true; there is a difference between the two formulations. If I express the gravitational potential as $\Sigma(M/R)$, then that is the answer. If I do not make any mistakes in my calculation, and do not forget any particles, I have got it right. On the other hand, if I know what the density everywhere is and thus know the Laplacian of the potential everywhere, then I have only got a second-order differential equation for the potential, and I have to have *boundary conditions* to specify it fully. Boundary conditions get us again on to this tiresome subject of cosmology, of what things are like far away, and we are back where we started.

Now what you can say if you go over to the relativistic case is that the usual field equations of general relativity are strictly parallel to the Laplacian formulation of Newtonian theory. They are second-order differential equations. Again you have to have boundary conditions of some kind or other to deal with the problem. Of course these are hyperbolic equations and you have a cone of influence and all that, but this does not help much. If, on the other hand, we had a formulation of the $\Sigma(M/R)$ kind, then we would have the complete answer; we would have the field without these tiresome questions of boundary conditions. What you want, then, is an action-at-a-distance or, if you like, particle formulation of general relativity. This is just what Hoyle and Narlikar tried to give.

And they were enormously encouraged in this by the fact that some 80 years after Maxwell formulated a field theory of electromagnetism, Wheeler and Feynman produced an action-at-a-distance formulation of electromagnetism. In other words, although for a long time it was thought that at least in the electromagnetic theory the field outlook was the only valid one, they could show that there was an equivalent way of formulating the subject in an action-at-a-distance manner. Now this was relatively easy for electromagnetic theory because electromagnetic theory has the wonderful property of being linear. And so you can really have something like $\Sigma(M/R)$ in electromagnetism. General relativity, as I pointed out earlier, is a non-linear theory. The whole problem of formulating an action-at-a-distance manner of describing a non-linear theory must be vastly more awkward than in the case of linear theory. So what Hoyle and Narlikar did was to write down a new formulation of gravitational theory. The matter is so involved that not all people agree with the authors that they have really got a pure action-at-a-distance formulation. This raises questions of taste and even of semantics. Hoyle and Narlikar submit their theory to the very severe test of the empty universe, which I do not myself regard as a useful test. They also indicate how, in the case of many particles, the theory goes over into general relativity. So here we have a new and highly interesting contender for the claim of linking Mach's principle with general relativity.

A theory of this degree of complexity needs a lot of examination to find its consequences in all likely and unlikely conditions. Of course, this will take time. Moreover, a theory which is unhappily so very untestable cannot really be adopted or failed except on grounds of taste. One might have hoped, as I was saying earlier, that a theory of Mach's principle would lead to a formulation in which one could test it, at least in principle, as opposed to some theory with an absolute space background or something of an equally irrelevant nature. But just because the Hoyle–Narlikar theory in a many-particle system leads to general relativity, I do not see any way—and I don't think Hoyle and Narlikar do—of discriminating, in the universe that we have got, between their theory and ordinary general relativity.

Let me conclude by discussing some other difficulties that are posed by our living in the universe we have got. The problems of cosmology are so awkward, from the philosophical and the practical scientific points of view, because our universe is so terribly unique. We come back here very much to what I said at the beginning of the chapter about physical experimentation, about the ability of switching forces on and off, or at least of having many different examples of motion, as in the solar system. But in the case of universe we have got precisely this one example. It is bound to affect our whole outlook enormously. When Galileo and Newton formulated gravitational theory, they naturally and reasonably and rightly formulated it by means of differential equations, because differential equations are a device for bringing a lot of different things under one hat. A differential equation is useful if it has the motions of the falling apple and of the moon and of the earth and of Jupiter and its satellites as solutions, because that shows that these are all phenomena of one kind. If we had anything more restricted than the Newtonian equations, we would not have a sufficiently wide framework to accommodate all the many examples we know of gravitation at work. But in the case of the universe we have just got one example. Therefore, why differentiate? We have got to take the *motion* of the universe, and not its *law of motion*. It is boring to describe separately the motion of the apple and of the moon and so on. But if there is nothing but one apple falling, then you would be silly if you did anything but describe that motion. So the best that we can hope to provide about the motion of the universe is a description, not a law of motion.

I should like to remind you of what I said earlier about the reformulation of Newton's law of inertia by general relativity so that one has certain specified motions that matter follows in the absence of forces. A force is then something that pulls matter off that standard. In the universe, what standard can we choose other than the actual motion? To ask, then, for the *forces* that are moving the universe is really asking for something irrelevant, since the best we can do is to describe the motion, and there is just one of it, and there is no general deviation from it. The universe moves as it does. The whole concept of force has no meaning in this context. It may be that much of what puzzles us about the structure of the universe is that we have not yet learnt to ask the right questions. It is not easy to think of how to ask them. It seems to me that some questions are clearly wrong ones, like: what force makes it move as it does? or: what would the universe be like if it were devoid of all matter? But saying that some questions are wrong does not tell us which are right. We shall have to learn a lot in this direction. The problem is, of course, that the universe cannot be shut off from our ordinary physics. It comes into it at every turn. It seems to me, as indeed to anybody who thinks there is something in Mach's principle, that the universe comes into every experiment because it provides the inertia of the bodies taking part in it.

Many of you may know that I have strong views about the importance of Olbers' paradox, about the fact that our expanding universe acts as a sink for light, that we send out much more radiation than we receive, and that the universe is very far from being in a state of thermodynamic equilibrium. Our whole attitude to any physics experiment is determined by this. In any experiment in which some energy is turned into radiation we expect it to travel far and to be lost. This whole mode of thinking is conditioned by our living in a universe with very particular properties. And so we are in the very uncomfortable situation that our differential equation physics has somewhere got to tie on to the descriptive physics that deals with the motion of the universe at large. But where local physics ends and cosmology begins is as indefinite and vague a question as any that I have discussed. This example of science not only being ignorant of the answers, but as yet even unable to formulate the important questions clearly, seems to mark an appropriate point to end this book.

SUGGESTIONS FOR FURTHER READING

Bondi, H. *Relativity and Common Sense*. London, Heinemann, 1964.

Bondi, H. 'Gravitational Waves', *Endeavour*, **20** (1961), 121–30.

Peierls, R. E. *The Laws of Nature*. London, Allen and Unwin, 1955.

Popper, K. R. *The Logic of Scientific Discovery*. London, Hutchinson, 1959.

Popper, K. R. *Conjectures and Refutations*. London, Routledge and Kegan Paul, 1963.

Sciama, D. W. *The Unity of the Universe*. London, Faber and Faber, 1959.

Weisskopff, V. F. *Knowledge and Wonder*. New York, Doubleday Anchor book, 1963.

INDEX

Index